LEXINGTON PUBLIC LIBRARY

Diplodocus

Lori Dittmer

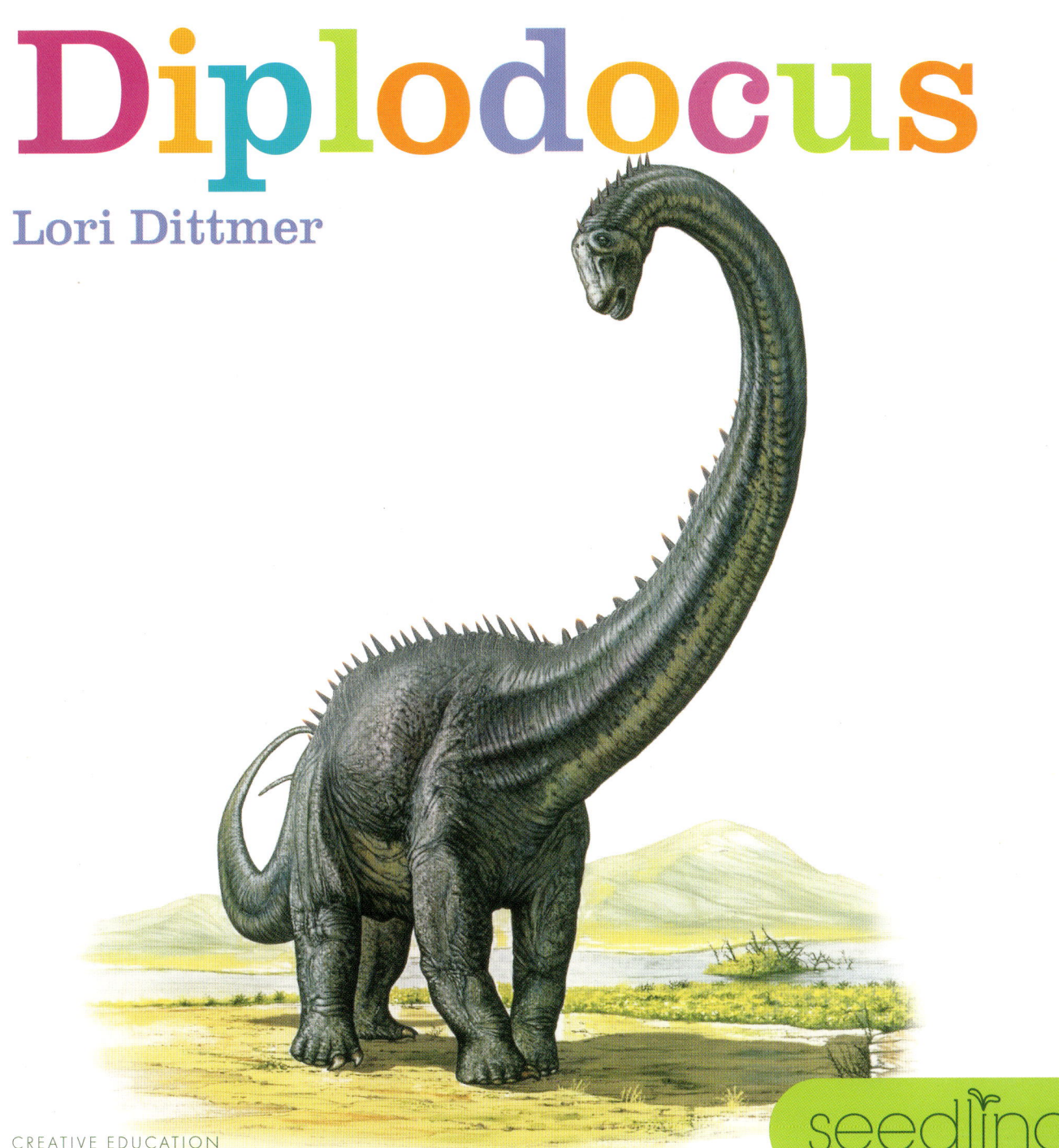

CREATIVE EDUCATION
CREATIVE PAPERBACKS

seedlings

Published by Creative Education and Creative Paperbacks
P.O. Box 227, Mankato, Minnesota 56002
Creative Education and Creative Paperbacks
are imprints of The Creative Company
www.thecreativecompany.us

Design by Ellen Huber
Production by Rachel Klimpel and Ciara Beitlich
Art direction by Rita Marshall

Photographs by Alamy (MasPix, The Natural History Museum, Science Photo Library, Universal Images Group North America LLC / DeAgostini), Dreamstime (Elena Duvernay), iStock (jondpatton), Science Source (Sebastian Kaulitzki, JAMES KUETHER), Shutterstock (Catmando, Herschel Hoffmeyer, I Wei Huang, SciePro, Vac1), Wikimedia Commons (Library of Congress Prints and Photographs Division)

Copyright © 2024 Creative Education, Creative Paperbacks
International copyright reserved in all countries.
No part of this book may be reproduced in any form
without written permission from the publisher.

Library of Congress Cataloging-in-Publication Data
Names: Dittmer, Lori, author.
Title: Diplodocus / by Lori Dittmer.
Description: Mankato, Minnesota : Creative Education and Creative Paperbacks, [2024] | Series: Seedlings: dinosaurs | Includes bibliographical references and index. | Audience: Ages 4–7 | Audience: Grades K–1 | Summary: "Early readers are introduced to Diplodocus, a Jurassic giant sauropod. Friendly text and dynamic photos share the dinosaur's looks, behaviors, and diet, based on scientific research"— Provided by publisher.
Identifiers: LCCN 2022013876 (print) | LCCN 2022013877 (ebook) | ISBN 9781640265028 (library binding) | ISBN 9781682770542 (paperback) | ISBN 9781640006324 (ebook)
Subjects: LCSH: Diplodocus—Juvenile literature. | Dinosaurs—Juvenile literature.
Classification: LCC QE862.S3 D584 2024 (print) | LCC QE862.S3 (ebook) | DDC 567.913—dc23/eng/20221026
LC record available at https://lccn.loc.gov/2022013876
LC ebook record available at https://lccn.loc.gov/2022013877

Printed in China

TABLE OF CONTENTS

Hello, *Diplodocus*! 4

Jurassic Dinosaurs 6

Early Discovery 8

Head to Tail 10

Claws and Spikes 12

What Did *Diplodocus* Do? 14

Plant Eaters 16

Goodbye, *Diplodocus*! 18

Picture a *Diplodocus* 20

Words to Know 22

Read More 23

Websites 23

Index 24

Hello, *Diplodocus!*

This dinosaur lived long ago.

Allosaurus and *Stegosaurus* lived then, too.

We know of *Diplodocus* because of its fossils.

8

It was discovered in 1877. O. C. Marsh named it.

This dinosaur was as long as three school buses.

It might have been able to stand up on its hind legs.

A large claw grew from each front thumb toe.

A row of spikes ran down its back.

Diplodocus walked on four legs.

It moved slowly. Its neck stretched for food.

Diplodocus ate plants.
Its teeth looked like pegs.

When an old tooth fell out, a new one grew in.

Goodbye, *Diplodocus!*

Picture a *Diplodocus*

Words to Know

discover: to find or learn of something for the first time

fossil: a bone or trace from an animal long ago that can be found in some rocks

hind: at the back

spike: a sharp, pointed object

Read More

Pimentel, Annette Bay. *Do You Really Want to Meet Diplodocus?* Mankato, Minn.: Amicus, 2020.

Sabelko, Rebecca. *Diplodocus*. Minneapolis: Bellwether Media, 2020.

Websites

DK Find Out! | *Diplodocus*
https://www.dkfindout.com/us/dinosaurs-and-prehistoric-life/dinosaurs/diplodocus
Read more about *Diplodocus* and take a dinosaur quiz.

PBS Eons | A Short Tale About *Diplodocus*' Long Neck
https://video.idahoptv.org/video/a-short-tale-about-diplodocus-long-neck-bvlquj
Watch a video to learn how *Diplodocus* and its relatives used their necks.

Note: Every effort has been made to ensure that the websites listed above are suitable for children, that they have educational value, and that they contain no inappropriate material. However, because of the nature of the Internet, it is impossible to guarantee that these sites will remain active indefinitely or that their contents will not be altered.

Index

claws, 12
discovery, 8, 9
feeding, 15, 16
fossils, 8
legs, 11, 14
length, 10
neck, 15
spikes, 13
teeth, 16, 17
when it lived, 6, 7